THE
COSTUME OF CHINA,

ILLUSTRATED IN

FORTY-EIGHT COLOURED ENGRAVINGS.

中国服饰

[英] 威廉·亚历山大 著

William Alexander

上海古籍出版社
SHANGHAI CHINESE CLASSICS PUBLISHING HOUSE

Selection of Rare Books from Bibliotheca Zi-ka-wei

徐家汇藏书楼珍稀文献选刊

徐锦华　主编

本书系国家社科基金重大项目
"徐家汇藏书楼珍稀文献整理与研究"
（项目批准号：18ZDA179）成果之一

总　序

董少新

文化交流是双向的，这是文化交流史研究的基本共识。但同时我们也必须认识到，在特定的历史时期，文化交流往往是不平衡的。这种不平衡体现在多个方面，其中就包括文化交流双方输入和输出的文化、知识、思想和物质产品等的数量不平衡，也包括己方文化对对方的影响程度不平衡。研究文化交流的这种不平衡性，考察特定历史时期文化交流双方输出和引进对方文化的数量及影响程度的差异，具有重要的学术意义。这样的研究可以在横向对比中为我们评估双方社会的发展程度、开放与包容性、对外来文化的态度、发展趋势及其原因等问题提供重要的参考。

16世纪以后的中西文化交流是人类历史上最伟大的文化交流之一。它不仅对双方造成了深刻的影响，而且一定程度上也促进了人类的近代化进程。对这一时期的中西文化交流史的研究，中外学界已有的成果可谓汗牛充栋。对前人研究略加梳理我们便不难发现，已有成果中更多的是研究西方文化东渐及其对中国的影响，而对中国文化西传欧洲的历史，尤其是中国文化在欧洲的影响史，虽然也有不少研究，但整体而言仍是远远不够的。这便给我们造成这样一种印象，认为西方先进的科技文化对中国造成了深远影响，而中国落后的农耕文化对西方的输出和影响十分有限；西方带动并主导了近代化进程，中国一度因为闭关锁国而错失了跟上先进的西欧发展步伐，最后不得不在西方的坚船利炮压力下才被迫打开国门，进入世界。导致这样的认知状况的原因很多，也很复杂，其中的一个重要原因是后见之明的影响，即用19世纪中西关系的经验来涵括整个16-20世纪中西关系史。如果我们以公元1800年为大约的分界线，将16世纪以来的中西关系史分为前后两个时期，那么不难看出，很多以往的观点和印象对16-18世纪的中西关系史并不适用，有的甚至是截然相反。

耿昇在法国学者毕诺（Virgile Pinot）《中国对法国哲学思想形成的影响》中译本"译者的话"中说："提起中西哲学思想和科学文化的交流，人们会情不自禁地想到西方对中国的影响。但在17-18世纪，中国对西方的影响可能要比西方对中国的影响大，这一点却很少有人提到。"在17-18世纪，到底是中国对西方的影响大，还是西方对中国的影响大？这是一个很值得思考并需要从多个角度加以回答的问题。

相关文献的数量或许是回答此问题的重要维度。就我个人的研究经验和观察而言，这一时期有关中国的西文文献的数量，要远远超过有关欧洲的中文文献数量。来华的西洋传教士、商人、使节和旅行家根据自己的所见所闻、亲身经历甚至中国典籍，用欧洲文字书写了数量庞大的书信、报告、著作和其他档案资料，绝大部分都被寄送或携带至欧洲，从而将丰富的中国信息传回了欧洲。这一时期曾到过中国的欧洲人数以万计，仅天主教传教士便有千余人，其中不乏长期在华、精通中文者。这些西方人是这一时期中西文化交流的主要媒介，其数量远超曾到过欧洲的中国人，而且黄嘉略、沈福宗、胡若望、黄遏东等少数去过欧洲的中国人，其主要扮演的角色和发挥的作用也是向欧洲传播中国文化和知识。来华传教士，尤其是实行适应性传教策略的耶稣会士，的确用中文翻译、撰写了数百部西学作品，但是数量上远不及他们以西文书写的介绍中国的书信、著作、报告乃至图册。也就是说，这一时期传入欧洲的中国知识和信息远多于传入中国的欧洲知识和信息。如果将带有丰富文化、艺术信息的瓷器、漆器、丝织品、外销画、壁纸、扇子等物质文化商品也考虑进来，中西文化交流在数量上的差距便更为明显，毕竟这一时期欧洲商人带入中国的作为商品的物质文化数量是相当有限的。

另一方面，17-18世纪欧洲的知识界根据来华传教士和商人带回的中国信息、知识而撰写的文章、小册子和书籍，其数量更远远超过中国知识界根据来华传教士和商人带至中国的欧洲信息、知识而撰写的作品。1587年出版的门多萨（Juan Gonsales de Mendoza）《中华大帝国史》（*Historia del Gran Reino de la China*）在欧洲被翻译成多种文字并一版再版，同时期没有一部有关欧洲的中国学者的作品出现；1735年出版的杜赫德（Jean Baptiste du Halde）《中华帝国全志》（*Description géographique, historique, chronologique, politique et physique de l'Empire de la Chine et de la Tartarie chinoise*）同样在欧洲广为流传，但同时期并没有一部中国学者的作品可与其相提并论，即便魏源的《海国图志》或可与之相比，其出版时间也已晚于《中华帝国全志》一个多世纪。从这个角度来看，中国知识和文化在欧洲的影响要远大于欧洲知识和文化在中国的影响。这一点还可以从欧洲盛行一个世纪的"中国风"以及一批启蒙思想家对中国文化的讨论中看出，而这一时期传入中国的欧洲艺术主要局限于清宫之中，从黄宗羲、顾炎武、王夫之到钱大昕、阎若璩、戴震等清代主流学界到底受到西学何种程度的影响，也还是不甚明了的问题。

当然，仅从文献数量来论证中国对欧洲有更大的影响并不充分。关于16-18世纪中国文化在欧洲的传播和影响，一个多世纪以来欧美学界有过不少专门的研究，如艾田蒲（Rene Etiemble）《中国之欧洲》、毕诺《中国对法国哲学思想形成的影响》和拉赫（Donald F. Lach）《欧洲形成中的亚洲》等。这些著作对打破欧洲中心主义的偏见发挥了重要作用，但我们也必须看到，这些著作在欧美学术界是边缘而非主流。在"正统"的欧洲近代史叙述中，欧洲所取得的成就是欧洲人的成就，是欧洲人对人类的贡献，是欧洲自古希腊、罗马时代以来发展的必然结果，包括中国在内的非欧洲世界的贡献及其对欧洲的影响几乎被完全忽视了。

中国学界方面，早在20世纪30-40年代，钱锺书、陈受颐、范存忠、朱谦之等学者便对中国文化在欧洲（尤其是英国）的传播和影响作了开拓性的研究。但此后中国学界对该问题的研究中断了较长的时间，直到20世纪90年代以来才重新受到学界的重视，出现了谈敏《法国重农学派学说的中国渊源》、孟华《伏尔泰与孔子》、许明龙《欧洲十八世纪中国热》、张西平《儒学西传欧洲研究导论：16-18世纪中学西传的轨迹与影响》、吴莉苇《当诺亚方舟遭遇伏羲神农：启蒙时代欧洲的中国上古史论争》、詹向红和张成权合著《中国文化在德国：从莱布尼茨时代到布莱希特时代》等一系列著作。但这些研究主要集中于中国文化对英、法、德三国启蒙思想家的影响，至于中国知识、思想、文化、物质文明、技术、制度等在整个欧洲的传播和影响这个大问题，仍有太多问题和方面未被触及，或者说研究得远非充分。例如，包括中国在内的非欧洲世界的传统知识、技术对欧洲近代科技发展有何种程度的影响？中国、日本、印度、土耳其乃至美洲的物质文化对欧洲社会风尚、习俗、日常生活的变迁起到了什么样的作用？近代欧洲逐渐形成的世俗化、宗教包容性、民主制度除了纵向地从欧洲历史上寻找根源之外，是否也存在横向的全球非欧洲区域的影响？世界近代化进程中，包括中国在内的非欧洲世界以何种方式发挥了怎样的作用？

对于这些问题的研究和讨论，首要的是掌握和分析16世纪以来欧洲向海外扩张过程中所形成的海量以欧洲语言书写的文献资料，其中就包括来华欧洲人所撰写的有关中国的文献，和未曾来华的欧洲人基于传至欧洲的中国信息和知识写成的西文中国文献。在这方面，西方学者比中国学者更有语言和文献学优势，在文献收集方面也拥有更为便利的条件。中国学界若要在中国文化西传欧洲及其影响问题上与欧美学界开展平等对话，乃至能够有所超越，必须首先在语言能力和文献掌握程度上接近或达到欧美学者的同等水准。实现这一目标极为不易，但近些年中国学界出现的一些可喜的变化，使我们对这一目标的实现充满期待，这些变化包括：第一，中西学界的交流越来越频繁和深入；第二，越来越多的年轻学者有留学欧美的经历，掌握一种乃至多种欧洲语言，并对近代欧洲文献有一定程度的了解，具备利用原始文献开展具体问题研究的能力；第三，中国学界、馆藏界和出版界积极推动与中国有关的西文文献的翻译出版，甚至原版影印出版。

上海图书馆徐家汇藏书楼拥有丰富的西文文献馆藏，不仅包括法国耶稣会的旧藏，而且包括近些年购入的瑞典汉学家罗闻达（Björn Löwendahl）藏书。徐家汇藏书楼计划从其馆藏中挑选一批珍贵的西文中国文献影印出版，以方便中国学界的使用。第一批出版的《中国植物志》《中华和印度植物图谱》《中国昆虫志》《中国的建筑、家具、服饰、机械和器皿之设计》《中国建筑》《中国服饰》均为17-19世纪初中西文化交流的重要文本或图册，是研究中国传统动植物知识、建筑、服饰、家具设计等在欧洲的传播和影响的第一手资料。这批西文文献，以及徐家汇藏书楼所藏的其他珍稀西文文献的陆续出版，无疑将推动中国学界在中学西传、中国文化对欧洲的影响等方面的研究。

（作者为复旦大学文史研究院研究员，博士生导师）

导　言

陆丹妮

地理大发现以来，东西方交流与日俱增，来自中国的丝绸、茶叶、瓷器源源不断地传入西方，西人对中国充满了向往，他们想要了解这个遥远、神秘的国度。旅人的东方游记以及传教士们关于中国的著作成为了他们认识中国的窗口，从《马可·波罗游记》（*The Travels of Marco Polo*）、门多萨（Juan Gonsales de Mendoza）的《中华大帝国史》（*Historia del Gran Reino de la China*）、利玛窦（Matteo Ricci）的《利玛窦中国札记》（*De Christiana expeditione apud Sinas*）、基歇尔（Athanasius Kircher）的《中国图说》（*China Illustrata*）到杜赫德（Jean Baptiste du Halde）的《中华帝国全志》（*Description géographique, historique, chronologique, politique et physique de l'Empire de la Chine et de la Tartarie chinoise*）等等，描绘中国的作品不胜枚举，其中关于中国题材的图画更是直观地向西人展现了来自东方的魅力。在画中可以看到充满异域风情的东方园林、繁荣的港口、穿着长袍马褂的中国人。若要对这些画作进行整理研究，必然绕不开一位英国画家——威廉·亚历山大（William Alexander）。

威廉·亚历山大出生于英国南部的肯特郡，自幼学习绘画。1784年2月，威廉·亚历山大在皇家美术学院正式注册为绘画专业的学生，1791年毕业。之后经过老师伊博森（Julius Caesar Ibbetson）的推荐，加入马戛尔尼使团前往中国。使团的正式画家另有其人，托马斯·希基（Thomas Hickey）是使团的官方正式画家（Painter），而威廉·亚历山大实则为托马斯·希基的助手，是一名制图员（Draughtsman）。不过，托马斯·希基留下的画作数量远远不及威廉·亚历山大。在访华期间，作为制图员的威廉·亚历山大随团访问了北京、杭州、广州、澳门等多地，创作了多幅水彩画和速写：从身着官服的清朝官员、拿着武器的士兵到雄伟高大的船舶，题材涉及中国的自然风光、军队士兵、城市建筑以及社会文化等多方面。

访华结束，回到英国的威廉·亚历山大出版了《中国服饰》（*The Costume of China, illustrated in forty-eight coloured engrarings*）与《中国人的服饰与习俗图鉴》（*Picturesque Representation of the Dress and Manners of the Chinese*）。展现在我们眼前的这本图册便是《中国服饰》。此书共收录了48幅经过

手工上色的铜版画，附有文字说明；题名页左侧另有一幅没有文字说明的《中国官员图》。从身穿清朝官服的官员到手握烟杆的清朝妇人，从繁忙的港口到高大巍峨的城墙，这些描绘着清朝人物形象、社会风俗、城市风情的铜版画向西方展示了中国的社会风情，使欧洲更好地了解中国，认识中国。

作为使团的一员，访华期间，威廉·亚历山大除了描绘日常见到的清朝官员、宫殿、士兵之外，还创作了多幅展现清朝普通百姓生活的作品。在此书，你可以看到身着朝服的官员、养尊处优的清朝妇人，也可以看到围坐在一起吃饭的纤夫、穿着蓑衣避雨的百姓。

《身披雨具的中国人》展现的就是清朝百姓穿着蓑衣、打着油纸伞的场景。画中一名男子蹲坐在路边，手上拿着一把油纸伞，背后的孩童躲在伞下避雨。另一名男子手拿烟杆，蹲坐在侧，身后一名男子穿着蓑衣避雨。威廉·亚历山大写道："在雨季，中国人会穿一件衣服（指蓑衣），这件衣服经过精心的设计，能让人们保持干燥，并且在很大程度上能够防止因暴露在潮湿环境中而产生的疾病。船工、农民和其他在户外工作的人，一般都有一件稻草制成的外套，雨水从上面流下，就像从水鸟的羽毛上流下一样。除此之外，他们有时还穿一件斗篷……完全覆盖在肩膀上；还有一顶由稻草和竹子组成的宽帽子，既能防晒又能防雨。如此装备的中国人，就像这幅画一样，可以抵御最猛烈的阵雨。"

此外，《中国服饰》一书还有不少展现中国城市建筑的作品。虽然威廉·亚历山大没能前往热河觐见乾隆皇帝，但经过协商，马戛尔尼使团"被允许经陆路从北京回到广州。他们乘坐平底帆船沿着白河和大运河南下，经过大约一个月的时间到达了大运河南面的终点杭州"。后来威廉·亚历山大转从舟山搭船从海路返回广州，没能领略从杭州南下广州的陆路风情，但是这段沿着大运河一路到杭州的经历让威廉·亚历山大得以尽情地施展他的才华，创作了不少反映当地社会风情、城市建筑的作品，这些作品部分被收录到了《中国服饰》中，例如《苏州附近的桥景》《杭州府附近的墓地》《牌楼》《定海塔》《舟山港：定海的南门》等等。其中《苏州附近的桥景》实际上描绘了大运河苏州段的景象，亚历山大写道："中国的桥梁构造各异。有许多是三层拱门，其中一些非常轻巧，而且很优雅；其他的一些只是金字塔形的桥墩，木头和地板水平铺在上面……桥梁的材料是一种粗糙的大理石，拱门上的五个圆形徽章上有中国文字，可能显示了建筑师的名字和建造日期。"《舟山港：定海的南门》描绘的则是定海的城墙："包围这座城市的城墙接近30英尺高，除了高塔及公共建筑，完全看不到那些只有一层的房屋……城墙上没有火炮，但在城墙上有供弓箭手使用的孔洞。在城墙上和城门的入口处，有帐篷作为岗哨，那里有足够数量的士兵驻扎。"值得注意的是，由于使团安排，威廉·亚历山大并未前往热河行宫觐见乾隆皇帝，而且返程到达杭州后转从舟山搭船由海路返回广州，因此部分画作是根据使团其他成员的速写或口述而创作的，难免有所纰漏。但我们也可借此了解当时西方人对中国的想象，认识他们眼中的中国。

马戛尔尼使团出访中国代表着东西方两个世界的撞击，是中西文化交流史上具有划时代意义的大事件，而作为随团成员创作的关于中国题材的绘画作品，自然吸引了不少人的眼光。1805年，《中国服饰》由印刷商威廉·米勒（William Miller）在伦敦出版，此书一经出版，便风行一时。《中国服饰》以及威廉·亚历山大笔下其他中国题材的作品成为了欧洲人了解中国社会风土人情的一个窗口。这些描绘中国题材的画作被不少书籍转载引用，引领了当时欧洲关于中国题材的创作风尚。19世纪英国维多利亚风格的画家托马斯·阿罗姆（Thomas Allom）在创作中国题材的作品时，便模仿借鉴了此书。阿罗姆从未到过中国，但他创作了诸多关于中国题材的绘画作品。《中国：那个古代帝国的风景、建筑和社会习俗》（*China, The Scenery, Architecture, and Social Habits of That Ancient Empire*）一书中的部分画作便借鉴模仿了威廉·亚历山大的作品。1821年完工的英国布莱顿皇家行宫，音乐厅墙上壁画中的建筑与人物均取自《中国服装》和亚历山大为斯当东那本书所绘制的对开本插图。

国内关于《中国服饰》的研究，大多是将其置于对威廉·亚历山大的研究之下，将《中国服饰》与《中国人的服饰与习俗图鉴》以及威廉·亚历山大为使团其他成员的著作所画插图结合起来进行整理研究。我们所据的《中国服饰》来自于上海图书馆徐家汇藏书楼的"罗氏藏书"，此藏本为小牛皮面装帧，竹节书脊烫金。扉页上有一句寄语:"感谢艾米丽·玛格丽特·谢弗的照顾,弟弟威廉。(Emily Margaret Sheaffe from her much affect. Brother William.)"题名页右上角有题词。此藏本质量颇高,细节丰富,画面充实饱满,色彩丰富。

我们将此书影印出版,希望透过威廉·亚历山大的画笔一探西方人眼中的中国,让读者领略18世纪末19世纪初他们心中那个奇幻瑰丽的"中国印象"。

(作者为上海图书馆历史文献中心助理馆员)

THE COSTUME OF CHINA

Emily Margaret Sheaffe.
from her much affect'e Brother William —

中国服饰

Published May 4 1805 by W. Miller, Old Bond Street, London.

THE

COSTUME OF CHINA,

ILLUSTRATED IN

FORTY-EIGHT COLOURED ENGRAVINGS.

BY
WILLIAM ALEXANDER.

LONDON:
PUBLISHED BY WILLIAM MILLER, ALBEMARLE STREET.
1805.

PORTRAIT OF VAN-TA-ZHIN,

A military Mandarine (or Nobleman) of China.

This officer (a colleague of Chow-ta-zhin, who was a mandarine of the civil department) was appointed by the Emperor to attend the British Embassy, from the time of its arrival in the gulf of Pe-tchi-li, till its departure from Canton. Van-ta-zhin was a man of a bold, generous, and amiable character, and possessed of qualifications eminently suited to his profession, being well skilled in the use of the bow, and in the management of the sabre. For services performed in the wars of Thibet, he wore appended from his cap, a peacock's feather, as an extraordinary mark of favour from his sovereign, besides a red globe of coral which distinguished his rank. He is represented in his usual, or undress, consisting of a short loose jacket of fine cotton, and an under vest of embroidered silk; from his girdle hang suspended his handkerchief, his knife and chopsticks* in a case, and purses for tobacco: on his thumbs are two broad rings of agate, for the purpose of drawing the bowstring. The heads of the arrows, which are thrust into the quiver, are variously pointed, as barbed, lozenge-headed, &c. His boots are of satin, with thick soles of paper: these are always worn by the mandarines and superior Chinese.

* Quoit-zau, or Chopsticks, are used in China instead of forks; they are two round slender sticks of ivory, ebony, &c. and used in the manner of pincers.

A PEASANT,

With his Wife and Family.

SMOKING tobacco is so universally prevalent in China, that it is not unusual to see girls of only twelve years of age enjoying this recreation. The Mother is in the dress of the northern provinces; the peak on her forehead is of velvet, and adorned with a bead of agate or glass. The hair is combed back so smooth by the assistance of oil, that it more resembles japan than hair; on the back of her head is a loop of leather, and the whole is kept together by bodkins of ivory or tortoise-shell. The general dress of this class of people, male or female, is nankeen dyed of various colours, though blue or black is most commonly worn.

The usual method of carrying infants, by mothers who are employed in any manufacture, or at any manual labour, as sculling of boats, &c. is by attaching them to the back in a kind of bag. Sometimes two children are seen fastened at the shoulders in the same manner. The Father wears appended from his girdle, a tobacco purse, knife case, and his flint and steel, by which the Chinese light a pipe very expeditiously. The elder Girl has her hair twisted into a hard knob at the crown, and ornamented with artificial flowers; she is prepared for dinner, having her bowl of rice by her, and her chopsticks in her hand. The feet of children are prevented from growing larger, by hard bandages bound strongly round them, the four smaller toes are turned under the foot, closely compressed, and the great toe forms the point. In consequence of this extraordinary custom the feet of adult women seldom exceed five inches and a half; even the peasantry pique themselves on the smallness of their feet, and take great care to adorn them with embroidered silk shoes, and bands for the ankles, while the rest of their habiliments display the most abject poverty.

A PAGODA (OR TOWER)

Near the City of Sou-tcheou.

THESE buildings are a striking feature on the face of the country. The Chinese name for them is Ta; but Europeans have improperly denominated them Pagodas, a term used in some Oriental countries for a temple of religious worship. It seems the Ta of China is not intended for sacred purposes, but erected occasionally by viceroys or rich mandarines, either for the gratification of personal vanity, or with the idea of transmitting a name to posterity; or perhaps built by the magistracy merely as objects to enrich the landscape.

They are generally built of brick, and sometimes cased with porcelain, and chiefly consist of nine, though some have only seven or five stories, each having a gallery, which may be entered from the windows, and a projecting roof, covered with tiles of a rich yellow colour, highly glazed, which receive from the sun a splendour equal to burnished gold. At each angle of the roofs a light bell is suspended, which is rung by the force of the wind, and produces a jingling not altogether unpleasant. These buildings are for the most part octagonal, though some few are hexagonal, and round. They diminish gradually in circumference from the foundation to the summit, and have a staircase within, by which they ascend to the upper story. In height they are generally from an hundred to an hundred and fifty feet, and are situated indiscriminately on eminences or plains, or oftener in cities. The Print represents one of modern structure. Those of a more ancient date are in a mutilated state, and the roofs covered with grey tiles, overgrown with moss, while others have a cornice only instead of the projecting roof.

Vide the print of Lin-tsin Pagoda in Sir George Staunton's Account of the Chinese Embassy.

中　国　服　饰

London. Publish'd July 20. 1797. by G. Nicol. Pallmall.

THE TRAVELLING BARGE

Of Van-ta-zhin.

As travelling in China is generally performed on the water, a prodigious number of Yachts or Barges of various forms are employed, as well for that purpose, as for the conveyance of merchandize.

The central apartment, which has an awning over the windows, is occupied by the proprietor; the fore part of the vessel by his servants, and the aft or stern part is used for culinary purposes, and sleeping places for the boatmen. Barges of this kind have one large sail of matting, stretched out by bamboos, running horizontally across it; the sail may be instantly taken in by letting go the haulyards, when the sail falls in folds similar to a fan. When the wind or tide is unfavourable, these vessels are either tracked along by human labour, or sculled by large oars which work on pivots at the bows and stern: by means of these oars, which are never taken out of the water, but simply sculled to and fro, the vessel is impelled onwards with considerable rapidity. The triple umbrella proclaims a Mandarine of consequence to be on board. The large lanterns with Chinese characters on them, and the ensign at the stern, are likewise marks of distinction.

12 中国服饰

London. Published October 12th 1797 by G. Nicol, Pall mall.

A CHINESE SOLDIER OF INFANTRY,

Or Tiger of War.

The dress of the Chinese is generally loose; the soldiers of this part of the army, with few exceptions, are the only natives whose close habit discovers the formation of the limbs.

The general uniform of the Chinese troops is cumbrous and inconvenient; this of the Tiger of War, is much better adapted for military action.

The Missionaries have denominated them Tigers of War, from their dress, which has some resemblance to that animal; being striped, and having ears on the cap.

They are armed with a scimitar of rude workmanship, and a shield of wicker or basket-work, so well manufactured, as to resist the heaviest blow from a sword. On it is painted the face of an imaginary monster, which (like that of Medusa) is supposed to possess the power of petrifying the beholder.

At a distance is seen a Military Post, with the Imperial flag, which is yellow, hoisted near it.

A GROUP OF TRACKERS

Of the Vessels, at Dinner.

When the wind or tide is unfavourable to the progress of the vessels, the sail and oars are laid aside, and the more general mode of tracking them is adopted. The number of trackers employed, depends on the size of the vessel, or strength of the current, which often requires the efforts of twenty men to counteract: these are kept in full exertion by a task-master, who most liberally applies the whip, where he sees a disposition to idleness.

The chief food of these poor labourers, is rice; and they consider it a luxury, when they can procure vegetables fried in rancid oil, or animal offal, to mix with it. They are represented cooking their meal over an earthen stove; the standing figure is employed eating his rice in the usual way, which is by placing the edge of the bowl against his lower lip, and with the chopsticks knocking the contents into his mouth.

They sometimes wear shoes made of straw, but are more frequently without any. The pien-za, or queue, is often inconvenient to Chinese labourers; to avoid which they twist it round their heads, and secure it by tucking in its extremity.

The flat boards, with cordage to them, are applied to the breast when dragging the junks, or vessels.

VIEW OF A BRIDGE,

In the Environs of the City of Sou-tcheou.

The Bridges of China are variously constructed. There are many of three arches, some of which are very light, and elegant; others are simply pyramidal piers, with timbers and flooring laid horizontally across them.

This arch, which resembles the outline of a horseshoe, occurred very frequently in the route of a part of the Embassy from Han-tcheou to Chusan. Like most of the Chinese bridges, it is of quick ascent, making an angle of full twenty degrees with the horizon, and is ascended by steps. The carriage of merchandize by land, is therefore inconsiderable; the rivers and canals being the high roads of China.

The material of which these bridges are composed, is a species of coarse marble. The projecting stones and uprights against the surface, are supposed to strengthen or bind the fabric; and the five circular badges over the arch, contain Chinese characters, which may probably shew the name of the architect, and date of its erection.

The temporary ornament over the centre of the arch, consisting of upright poles, painted and adorned with silken streamers, and suspended lanterns, was erected in compliment to the Embassador. The six soldiers from an adjacent Military Post, were likewise ordered to stand on the bridge, by way of salute.

中 国 服 饰

London Publish'd, Oct.r 12, 1797, by G. Nicol, Pallmall.

PORTRAIT OF A TRADING SHIP.

These ships venture as far as Manilla, Japan, and even Batavia, which is the most distant port they visit; and many of them are from eight hundred to a thousand tons burthen. In these voyages the mariners take the moderate season of the year, and though well acquainted with the use of the compass, generally keep near the coast.

No alteration has been made in the naval architecture of China for many centuries past. The Chinese are so averse to innovation, and so attached to ancient prejudices, that although Canton is annually frequented by the ships of various European nations, whose superiority of construction they must acknowledge, yet they reject any improvement in their vessels.

The stern of this ship falls in with an angle; other vessels are formed with a cavity, in which the rudder is defended from the violence of the sea; yet this contrivance certainly subjects the ship to much hazard, when running before the wind in high seas.

On each bow is painted an eye, with the pupil turned forwards; perhaps with the idea of keeping up some resemblance to a fish; or from a superstitious notion, that the ship may thus see before her, and avoid danger.

The ports often serve as windows, not many of them being furnished with ordnance.

PORTRAIT OF THE PURVEYOR

For the Embassy, while the Embassador remained at Macao.

The dress of this figure is the same as is generally worn by the citizens, or middle class of people in China, with variations in the colour; and some difference of form in hats, caps, boots, &c. &c.

The external jacket is of sheep-skin, ornamented with crescents of the same material, dyed of another colour, sewed into it at equal distances; and has a collar of sable, or fox skin. This surtout is worn on such mornings and evenings as are fresh and cold; in the day time (if found inconveniently hot) it is laid aside. Under this is worn a vest of figured silk; beneath which is another of white linen, or taffeta; and lastly, a pair of loose drawers: in the summer season these are of linen or silk, and for the winter, they are lined with fur, or quilted with raw silk; and in the northern provinces they are worn, made of skins only.

The cap is composed of a coarse sort of felt, which is very common; and while new, they have the shape of those worn by the Mandarins, (see the Portrait of Van-ta-zhin,) but they soon become pliant and misshapen, by wear, or when rain has taken the stiffness from them. The stockings are of nankeen, quilted on the inside with cotton. The shoes are likewise nankeen, with thick soles made of paper.

From the girdle on the right side, hangs a flint and steel, and knife sheath; on the left, purses for tobacco, or snuff.

The box held in his hand contains sweetmeats; a jar of which he entreated the persons of the Embassy to accept as a token of his regard.

The back ground is a scene at Macao.

中 国 服 饰

PUNISHMENT OF THE CANGUE,

By which name it is commonly known to Europeans, but by the Chinese called the Tcha; being a heavy tablet, or collar of wood, with a hole through the centre, or rather two pieces of wood hollowed in the middle, which inclose the neck (similar to our pillory); there are, likewise, two other holes, for the hands of the delinquent, who is sometimes so far favoured as to have but one hand confined; by which indulgence he is enabled with the other to lessen the weight on his shoulders.

The division in the Cangue which receives the head, is kept together by pegs, and is further secured by a slip of paper pasted over the joint, on which is affixed the seal, or chop, of the Mandarin; and the cause of punishment likewise depicted on it, in large characters.

The weight of these ignominious machines, which are from sixty to two hundred pounds in weight, and the time criminals are sentenced to endure them, depends on the magnitude of the offence, being sometimes extended, without intermission, to the space of one, two, or even three months; during which time the offender's nights are spent in the prison, and in the morning he is brought by the magistrates' assistant, led by a chain, to a gate of the city, or any place most frequented; when the attendant suffers him to rest his burthen against a wall, where he remains exposed throughout the day to the derision of the populace, without the means of taking food but by assistance. Nor is the punishment at an end when the Mandarin has ordered him to be released from the Cangue; a certain number of blows from the bamboo, remain to be inflicted; for which chastisement, in the most abject manner, with forehead to the earth, he thanks the Mandarin for his fatherly correction.

SOUTH GATE OF THE CITY OF TING-HAI,

In the Harbour of Tchu-san.

THE Port of Tchu-san, into which the English were formerly admitted, lies in latitude, thirty degrees and twenty minutes north, or about midway, on the east coast of China, between Can-ton and Pe-king.

The walls inclosing this city are near thirty feet in height, which (excepting Pagodas, public buildings, &c.) entirely preclude the sight of the houses, which in general have but one story.

The bricks and tiles of China, either from a different quality of the substance that composes them, or from being dried and burnt in a different manner, are of a bluish, or slate colour. The embrasures have no artillery, but there are loop-holes in the merlons for the use of archers. On the walls, and at the entrance of the gate, are tents as guard-houses, where a sufficient number of soldiers are continually stationed. At an early hour of the night the gates are shut, after which no person can be admitted on any pretence whatever.

The angles of the roofs which curve upwards, and project considerably, in Chinese buildings, most likely have their origin from tents; for a canvas resting on four cords will receive the same form. The ridges on the angles of the buildings over the gate are decorated with figures of animals, dragons, &c.; and the sides of the building, and extremities of the beams, painted with various colours. The yellow board over the arch has Chinese characters on it, which probably signify the name and rank of the city. The carriage entering the city, is a vehicle used in common with sedans, for the conveyance of persons of consequence. The Chinese have not adopted the use of springs, therefore these machines are little better than a European cart. The nearest figure shews the usual method of carrying light burthens, as vegetables, fruit, &c. &c.

THREE VESSELS LYING AT ANCHOR

In the River of Ning-po.

The middle vessel, with the stern in view, was a trading ship without cargo; in this the peculiar construction of the stern is exemplified, being hollowed into an indented angle, for the protection of the rudder, which is lifted out of the water by a rope, to preserve it. The Chinese characters over the rudder, denote the name of the vessel; and the bisected cone against the stern, is appropriated to the same use as the quarter-galleries of our ships.

The small vessel was hired for the service of the Embassy, and employed in transporting baggage; the larger vessel conveyed a part of the Embassy from Ning-po, to Tchu-san, where they embarked on board the Hindostan, for Can-ton. The prow of this vessel has a singular appearance, the upper part of the stern terminating in two wings, or horns. The small boat (or Sam-paan, as called by the Chinese) is a necessary appendage to vessels of this size.

中 国 服 饰

PORTRAIT OF A LAMA, OR BONZE.

The priesthood of China and Tartary are, since the conquest of the former, become nearly the same, in respect to manners, dress, &c.; and these are the only people of either nation, who have the head shaved entirely. Their general habit is a loose robe or gown, with a broad collar of silk or velvet; the colour of the robe depending on the particular sect or monastery to which they belong. Some of them wear an ornament resembling a cap, exquisitely wrought in wood, &c. which they affix to the back of the head.

This figure is from one of the Lamas inhabiting the temple called Poo-ta-la, which is situated near the Imperial residence at Zhe-hol in Tartary. These Priests are all clad in the royal colour, yellow; their hats have very broad brims, answering the double purpose of defence from sun and rain, and are neatly manufactured from straw and split bamboo.

The temple Poo-ta-la, which is distantly seen, maintains eight hundred Lamas, devoted to the worship of the deity Fo: to this sect the Emperor is attached, and it is the general religion of the empire. The form of this edifice is square, with lesser buildings in the Chinese style of architecture adjoining: each side of the large building measures two hundred feet, and is nearly of the same height, having eleven rows of windows. In the centre of this immense fabric is a chapel, profusely decorated and roofed with tiles of solid gold. Within this chapel is the sanctum sanctorum, containing statues of the idol Fo, with his wife and child.

A CHINESE LADY AND HER SON,

attended by a Servant.

THE female sex in China live retired in proportion to their situation in life. The lower orders are not more domesticated than in Europe; but the middle class are not often seen from home, and ladies of rank scarcely ever. Alterations of dress are never made from caprice or fashion; the season of the year, and disposing the various ornaments, making the only difference. Instead of linen, the ladies substitute silk netting; over which is worn an under vest and drawers of taffeta; and, (should the weather require no additional covering,) they have for the external garment, a long robe of silk or satin, richly embroidered. Great care is taken in ornamenting the head: the hair, after being smoothed with oil and closely twisted, is brought to the crown of the head, and fastened with bodkins of gold and silver; across the forehead is a band, from which descends a peak of velvet, decorated with a diamond or pearl, and artificial flowers are fancifully arranged on each side of the head. Ear-rings, and the string of perfumed beads suspended from the shoulder, likewise make up part of the ornaments of dress. The use of cosmetics is well known among the ladies of China; painting the face both white and red, is in common practice with them: they place a decided red spot on the lower lip, and the eyebrows are brought by art to be very narrow, black, and arched.

The small shoes are elegantly wrought, and the contour of the ankles are never seen, by reason of the loose bandage round them. Boys, till about seven years of age, frequently have two queues, encouraged to grow from each side of the head. The servant, as is usual with the lower class, wears on the wrist a ring of brass or tutenag.

VIEW OF A BURYING-PLACE,

near Han-tcheou-fou.

THE tombs and monuments of China exhibit a variety of architecture, except those of the common people, which are nothing more than small cones of earth, on the summits of which they frequently plant dwarf trees. These simple graves are occasionally visited by the family, who are particularly careful to trim and keep them in neat order.

The coffins of this country are made of very thick boards, plentifully pitched within, and varnished without; which makes them durable, and prevents them from emitting putrid exhalations: this process being absolutely necessary, where the coffins of the lower class often lie scattered among the tombs, totally uncovered with earth.

The rich spare no expence in having coffins of the most precious wood, which are frequently provided several years before the death of the persons intending to occupy them. A deceased parent is oftentimes preserved in the house by an affectionate family for months, and even years; yet, either from their knowledge of embalming, or from the practice of securing the joints of the coffin with bitumen, no contagious effluvia proceeds from it.

The duty of the widow or children is not finished here; even after the corpse is deposited in the sepulchre of its ancestors, the disconsolate relatives (clad in coarse canvas) still reside with the body, and continue their lamentations for some months. The characters on the monuments, signify the name and quality of the defunct; and epitaphs, extolling the virtues of the deceased, are inscribed on tablets of marble at the entrance of the vaults. The tomb with steps before it, and another, inclosed with cypresses, are common with people of affluence.

FRONT VIEW OF A BOAT,

passing over an inclined Plane or Glacis.

In the passage from Han-tcheou-fou to Tchu-san (which was the route of part of the Embassy), the face of the country is mountainous; therefore the communication of the canals is continued by means of this sort of locks, two of which were passed over on the 16th of November, 1793.

In this subject, the difference of level between the two canals was full six feet; in the higher one, the water was within one foot of the upper edge of the beam over which the boat passes. The machinery consisted of a double glacis of sloping masonry, with an inclination of about forty degrees from the horizon. The boats are drawn over by capstans, two of which are generally sufficient, though sometimes four or six are required for those of greater burden; in this case, there are holes in the ground to receive them. When a boat is ready to pass over, the ropes from the capstans (which have a loop at their extremities) are brought to the stern of the vessel; one loop is then passed through the other, and a billet of wood thrust into the noose, to prevent their separation; the projecting gunwale at the same time keeping the ropes in a proper situation. This being adjusted, the men heave at the capstans till the boat has passed the equilibrium, when, by its own gravity, it is launched with great velocity into the lower canal, and is prevented from shipping too much water, by a strong skreen of basket-work, which is placed at the head. On the left hand stands a mutilated triumphal arch, and a small temple inclosing an idol, to which sacrifices are frequently made for the preservation of the vessels passing over.

For a plan and section of the above, vide Sir George Staunton's Account, Plate 34 of the folio volume.

中 国 服 饰

PORTRAIT OF A SOLDIER,

in his full Uniform.

THE empire of China has, since the conquest of the Tartars, enjoyed uninterrupted tranquillity, if we except partial insurrections, &c. and in consequence of this long intermission of service, the Chinese army are become enervated, and want the courage, as well as the discipline, of European troops; for strict order is so little enforced, that it is not uncommon to see many among them fanning themselves while standing in the ranks.

The candidates for promotion, in their army, are required not only to give proofs of their knowledge in military tactics, but they must likewise exhibit trials of personal strength and agility, by shooting at the target, exercising the matchlock, sabre, &c.

The situation of the soldiery is even envied by the lower classes, as they regularly receive their pay, though their services are seldom required, but occasionally to assist in quelling tumults, or doing duty at the military posts; thus, for the greater part of their time, they follow their several occupations, having little else to do than keep their arms and accoutrements bright and and in good order, ready for the inspection of the officers, should they be suddenly called out to a review, or any other emergency.

This dress of the troops is clumsy, inconvenient, and inimical to the performance of military exercises, yet a battalion thus equipped has, at some distance, a splendid and even warlike appearance; but on closer inspection these coats of mail are found to be nothing more than quilted nankeen, enriched with thin plates of metal, surrounded with studs, which gives the *tout-ensemble* very much the appearance of armour.

From the crown of the helmet (which is the only part that is iron) issues a spear, inclosed with a tassel of dyed horse-hair. The characters on the breast-plate, denote the corps to which he belongs; and the box which is worn in front, serves to contain heads of arrows, bowstrings, &c. &c. The lower part of the bow is inclosed in a sheath or case.

中 国 服 饰

A GROUP OF PEASANTRY, WATERMEN, &c.

playing with Dice.

THE Chinese are so much addicted to gaming, that they are seldom without a pack of cards, or a set of dice. Cock-fighting is in practice among them; and quails are also bred for the same purpose. They have likewise a large species of grasshopper (or grillæ) common in China; a couple of these are put into a bason to fight, while the by-standers bet sums of money on the issue of the conflict: these insects assail each other with great animosity, frequently tearing off a limb by the violence of their attacks. The Chinese dice are marked exactly similar to those of Europe; in playing they never use a box, but cast them out of the hand. The laws of the empire allowing them full power to dispose of their wives and children, instances have happened when these have been put to the hazard of a throw; and it should be mentioned, that in all their games, whether for amusement or avarice, the Chinese are very noisy and quarrelsome. The figure standing with an instrument of agriculture in his hand, is an husbandman; another sitting figure, with a small black cap, is a waterman, having by him a gong, which is an instrument of semi-metal resembling a pot-lid; this being struck with the stick lying near it, produces a harsh jarring sound, which is heard at a considerable distance: one of these is always suspended at the head of every vessel when tracked along the canals, and struck as occasion requires, by the people on board, to inform the trackers when to desist hauling, and when to resume their labour. By this method much confusion is prevented, where the great concourse of vessels would be continually running foul of each other, if not warned by this contrivance.

These gongs have so many various notes, that the trackers know perfectly when the signal is made from the vessel they are hauling.

中 国 服 饰

VIEW OF A CASTLE,
near the City of Tien-sin.

This castle, or tower, is situated on a point of land at the confluence of three rivers, the Pei-ho, the Yun-leang, and the When-ho, near the celestial city (Tien-sin), which is the chief harbour for shipping, and principal depot for merchandize throughout China; and from whence the various articles of commerce are circulated, by means of the canals, through the most distant provinces.

This edifice is thirty-five feet in height, and built with bricks, except the foundation, which is of stone, and has been undermined, most likely by indundation; the surrounding country being very low and marshy. A guard of soldiers is constantly stationed here, and, in cases of tumult or commotion, the centinels give the alarm to the adjacent military posts, in the daytime by hoisting a signal, and at night by the explosion of fireworks; on which the neighbouring garrisons repair to the spot where their services are required.

Within the battlements is a building to shelter centinels on duty; one of them is beating a gong, to announce to the garrison the approach of a viceroy or mandarin of rank; on this notice, they immediately form in a rank, and stand under arms to salute him. Within the parapet a lantern is suspended, and in the opposite angle the imperial standard is elevated; the colour of the tablet, with the inscription on it, likewise shews it to be a royal edifice, In Nieu-hoff's account of the Dutch embassy, which was sent to Pekin in the year 1650, is a print either of this tower, or one similar to it, which stood on the same site. The hillocks of earth under a clump of trees, seen in the distance, are burying-places.

中 国 服 饰

A SEA VESSEL UNDER SAIL.

SHIPS of this construction are employed by the merchants, in conveying the produce of the several provinces to the different ports of the empire.

The hold for the stowage of the various commodities, is divided into several partitions, which are so well caulked, with a composition called chu-nam, as to be water-proof; by this contrivance, in the event of a leak, the greater part of the cargo is preserved from injury, and the danger of foundering considerably removed.

The main and foresails are of matting, strongly interwoven, and extended by spars of bamboo running horizontally across them; the mizen and topsails are nankeen, the latter of which is (contrary to the European method) never hoisted higher than is seen in the drawing. The sails are braced up or eased off, by means of ropes attached to the extremities of the spars in the sails, which are known by the name of a crowfoot; and thus the ship is tacked with very little trouble.

The prow, or head is, as usual with Chinese vessels, without stem; they are likewise without keel, and consequently make considerable leeway. The two anchors are made of a ponderous wood, called by the Chinese tye-mou, or iron wood, the several parts of which are strongly lashed and bolted together, and pointed with iron, though sometimes they carry large grapnels of four shanks. The arched roof of matting is the cabin, in which the seamen sleep, &c. and the bamboo spars on the quarter, are conveniently carried in that situation for the uses of the ship.

The several flags and ensigns are characteristic of the taste of the Chinese.

中 国 服 饰

PORTRAIT OF CHOW-TA-ZHIN,

In his Dress of Ceremony.

CHOW-TA-ZHIN, a Quan, or Mandarin, holding a civil employment in the state, was, with Van-ta-zhin, entrusted by the Emperor with the care of the British Embassy during its residence in China. He was a man of grave deportment, strict integrity, and sound judgment, as well as of great erudition; having been preceptor to a part of the Imperial family.

His external honours were the customary distinction of a blue ball on his cap; from which was suspended a peacock's feather, being a mark of additional rank.

He is attired in his full court dress, being a loose gown of silk or satin, covering an under vest richly embroidered in silk of the most vivid colours; the square badge on his breast, and its exact counterpart on the back, is also of rich embroidery, and contains the figure of an imaginary bird, which denotes the wearer to be a Mandarin of letters, in like manner as a tiger on the badge would shew the person to be in a military capacity. The beads worn round the neck are occasionally of coral, agate, or of perfumed wood, exquisitely carved, as affluence or fancy may dictate.

In his hand he holds a paper relative to the Embassy.

中 国 服 饰

A CHINESE PORTER, OR CARRIER.

When the wind is favourable, and where the level face of the country will admit, the Chinese sometimes hoist this simple kind of sail to lessen the exertion of the driver; when the wind is adverse, the sail is laid aside, and another labourer employed to assist in pulling the machine, by means of a rope placed across his shoulders.

The carriage contains, among other articles, some vegetables, a basket of fruit, a box of tea, loose bamboos, and a jar of wine, the stopper of which is covered with clay, to prevent the air injuring the liquor; on the side are placed his hat, and some implements for keeping the machine in order.

This contrivance is thus described by Milton, in his Paradise Lost, Book III. line 437, &c.

" But in his way lights, on the barren plains
" Of Sericana, where Chineses drive,
" With sails and wind, their cany waggons light."

中 国 服 饰

THE HABITATION OF A MANDARIN.

The house of a Mandarin is generally distinguished by two large poles erected before the gate; in the day-time flags are displayed on these poles as ensigns of his dignity, and during the night painted lanterns are suspended on them.

The superior Chinese choose to live in great privacy, their habitations therefore are generally surrounded by a wall; their houses seldom exceed one story in height, though there are some few exceptions, as in the residence of the Embassy at Pekin, where one of the many edifices of that palace had apartments above the ground floor, and was occupied by the Secretary of the Embassy.

The several rooms of a Chinese house are without ceilings, so that the timbers supporting the roof are exposed. The common articles of furniture are, frames covered with silk of various colours, adorned with moral sentences, written in characters of gold, which are hung in the compartments; on their tables are displayed curious dwarf trees, branches of agate, or gold and silver fish, all which are placed in handsome vessels of porcelain.

A MANDARIN'S TRAVELLING BOAT.

Mandarins, who are employed in travelling from place to place on the public service, keep barges for that purpose, as carriages are kept in England.

They are generally ornamented by painting and varnishing the pannels and mouldings with various devices, &c. At night, or during rain, the part occupied by the Mandarin is inclosed by shutters, and the light is then received through lattices, covered with laminæ of oyster shells.

The gunwale of these barges (as with most Chinese vessels) is sufficiently broad for the watermen, &c. to pass from stem to stern, without inconvenience to passengers in the principal apartments.

The Mandarin is seen attended by soldiers and servants, who are bringing his dinner; the double umbrella, or ensign of his authority, is conspicuously placed to demand respect; the flag and board at the stern, with Chinese characters on them, exhibit his rank and employment; these insignia of power also serve as a signal for other vessels to make clear passage for him, in consequence of which, such boats are seldom obstructed in their progress through the immense number of vessels constantly employed on the canals. The master of any vessel who, by mismanagement, or even accident, should impede these officers in the exercise of their duty, would most likely receive the instant punishment of a certain number of blows from the bamboo, at the discretion of the Mandarin.

A STANDARD BEARER.

Early in the morning of the 30th of September, 1793, the Embassador and suite proceeded on their journey northward, to pay the customary compliment of meeting the Emperor, who was then returning from his summer residence in Tartary, to his palace at Pekin; on this occasion, each side of the road was lined, as far as the eye could reach, with mandarines, soldiers, &c. bearing banners, large silk triple umbrellas, and other insignia of Chinese royalty. The Print represents a soldier employed in bearing a standard, or gilt board, on which are depicted characters, which probably display some title of the Emperor.

His dress is nankeen cotton, which is tied round the waist, with the imperial or yellow girdle, and his legs are cross-gartered: his hat is straw, neatly woven, and fastened under the chin; the crown is covered with a fringe of red silk, converging from the centre, where a feather is placed.

His sword, as is customary with the Chinese, is worn with the hilt behind.

A SACRIFICE AT THE TEMPLE.

The Chinese have no regular sabbath, or fixed time for worshipping the Deity in congregation. Their temples being constantly open, are visited by the supplicants on every important undertaking, such as an intended marriage, the commencement of a long journey, building a house, &c.

The figure on the right hand is anxiously watching the fall of tallies, which he is shaking in a joint of bamboo; these are severally marked with certain characters, and as they fall, the characters are inserted by the priest in the book of fate. After the ceremony, the priest communicates to the votary the success of his prayers, which has been thus determined by lot.

The priesthood always shave the head entirely, and wear a loose dress of silk or nankeen, the colour of which is characteristic of their particular sect.

The figure kneeling before the sacred urn, in which perfumed matches are burning, is about to perform a sacrifice. On these occasions round pieces of gilt and silvered paper are burnt in tripods for that purpose, and at the same time quantities of crackers are discharged.

Behind the figures are seen two hideous idols. These statues are usually arranged against the walls of the temple, inclosed within a railing.

A MILITARY STATION.

Along the canals and public roads of China, great numbers of military posts are erected, at which eight or ten soldiers are generally stationed.

Adjacent to each of these stands a look-out-house, commanding an extensive prospect; and adjoining are placed five cones of plastered brick work, out of which certain combustibles are said to be fired, in times of alarm from invasion or insurrection. In front of the building is a simple triumphal entrance, on which is an inscription suitable to the place. Near this the imperial ensign is elevated; and on the left of the house is a frame of wood, in which are deposited different arms, as pikes, matchlocks, bows, &c.

The vessel passing by with a double umbrella, contains some mandarin of distinction, who is saluted by the firing of three petards,* and by the guard, who are drawn out in a rank.

* The Chinese, on these occasions, never use more than three guns, which are always fired perpendicularly, to prevent accidents.

A FISHING BOAT.

This Print illustrates a contrivance of the Chinese fishermen for raising their nets: the frame work is composed of that most useful plant the bamboo, which, uniting strength with lightness, is made use of on almost every occasion. When the weight of a man at the extremity of the lever is insufficient to lift a large draught of fish, he is assisted by a companion, as in the representation; the rest of the company are employed at dinner, steering, &c. protected from the sun and weather by a rude covering of mats: the boat is also provided with grapnels, and a lantern to prevent accidents at night. The distance is a view of the lake Poo-yang. On the left hand, near the benches, are some mounds of earth, which occur occasionally for several miles together; the purpose generally assigned to them is the repairing any accidental breach of the canal, with all possible expedition.

Another mode of fishing, often practised by the Chinese, is by means of a species of pelican, called the Leu-tze. See the Account of the British Embassy, by Sir George Staunton, Vol. II. p. 388.

A CHINESE COMEDIAN.

Theatrical exhibitions form one of the chief amusements of the Chinese; for though no public theatre is licensed by the government, yet every Mandarin of rank has a stage erected in his house, for the performance of dramas, and his visitors are generally entertained by actors hired for the purpose.

On occasions of public rejoicing, as the commencement of a new year, the birth-day of the Emperor, and other festivals, plays are openly performed in the streets, throughout the day, and the strolling players rewarded by the voluntary contributions of the spectators.

While the Embassador and his suite were at Canton, theatrical representations were regularly exhibited at dinner time, for their diversion. This character, which the Interpreter explained to be an enraged military officer, was sketched from an actor performing his part before the embassy, December 19, 1793.

These entertainments are accompanied by music: during the performance of which, sudden bursts, from the harshest wind instruments, and the sonorous gong, frequently stun the ears of the audience.

Females are not allowed to perform: their characters are therefore sustained by eunuchs; who, having their feet closely bandaged, are not easily distinguished from women.

The dresses worn by players, are those of ancient times.

A GROUP OF CHINESE,

Habited for Rainy Weather.

During the rainy seasons, the natives of China wear an external dress, well calculated to keep them dry, and prevent, in a great measure, such diseases as arise from exposure to wet.

Watermen, peasantry, and others, employed in the open air, are generally provided with a coat made of straw, from which the rain runs off, as from the feathers of an aquatic bird: in addition to this, they sometimes wear a cloak, formed of the stalks of kow-liang (millet), which completely covers the shoulders; and a broad hat, composed of straw and split bamboo, which defends them both from sun and rain. A Chinese thus equipped as is the standing figure,) may certainly defy the heaviest showers.

The soldier, under an umbrella of oiled canvas, wears his undress; consisting of a jacket, of black nankeen, bordered with red; behind him is his child, to whom he is likewise affording shelter.

The figure smoking, is habited in a large coat, of skin, with the hair, or wool, remaining on it: sometimes the coat is turned, and the hairy side worn inwards.

中 国 服 饰

A PAGODA, OR TEMPLE,

For religious Worship.

The Chinese are scrupulously observant of moral and religious duties; and their country abounds with temples, of various forms, to which they resort, on every interesting occasion, and offer their sacrifices. Besides these temples, a small tabernacle, or niche, containing their household gods, is to be found in almost every house and ship.

Some religious ceremonies of the Chinese resemble those of the Church of Rome: and the Chinese Idol, denominated Shin-moo, is very similar to the representations of the Virgin and Child; both being figures of a female and an infant, with rays of glory issuing from their heads, and having lights burning before them, during the day as well as night.

The greater part of the people, are of the sect of Fo; whose followers believe in the metempsychosis, and in a future state of happiness, after a virtuous life; and suppose, that the souls of the irreligious live hereafter in a state of suffering, and subject to the hardships endured by inferior animals.

The figures dressed in loose gowns, are priests, attending at the temple; and the back ground, is a view of the city Tin-hai, Nov. 21, 1793.

中国服饰

A SHIP OF WAR.

The Chinese are so well supplied with the produce of their own country, as to require very little from distant lands; and it is to this native abundance the low state of navigation among them ought to be attributed.

Though they are said to have been acquainted with the use of the compass, from the earliest ages, yet they cannot be considered as expert seamen, either in their application of astronomy to nautical purposes, or skill in manœuvring their clumsy ships.

The compass is, however, an instrument venerated by the seamen, as a deity; and to which they sometimes offer sacrifices of flesh and fruit.

The drawing was made from a ship (Pin-gee-na) lying at anchor in the river, near Ning-po. These vessels may properly be termed floating garrisons; as they contain many soldiers, and are generally stationed near their principal towns.

These soldiers often hang their shields against the ship's quarter; and the rudder is lifted, by ropes, nearly out of the water, perhaps to preserve it, while at anchor.

The ports are false; as few ships of the Chinese navy are, at present, supplied with artillery.

中 国 服 饰

A SOLDIER IN HIS COMMON DRESS.

The army of China cannot be considered formidable, their troops being naturally effeminate, and without the courage of European soldiers: one reason assigned for this is a mode of education which is not calculated to inspire a nation with courage, and it may partly be accounted for, from their having enjoyed uninterrupted peace since their subjugation by the Tartars.

Every soldier, on his marriage, and on the birth of a male child, is intitled to a donation from the Emperor; and the family of a deceased soldier receives likewise a gift of condolence.

The undress of a Chinese or Tartar soldier consists of a short jacket of black or red nankeen, with a border of another colour; under this is a garment of the same material, with long sleeves: when the weather is cold, one or more dresses are worn under this. The flag at his back is of silk, and fastened by means of a socket attached behind: these are generally worn by every fifth man, and make a very gay appearance.

Their bows are of elastic wood, covered on the outside with a layer of horn, and require the power of from seventy to one hundred pounds in drawing them; the string is composed of silk threads closely woulded, and the arrows are well made and pointed with steel. Their scymeters, though rudely formed, are said to equal the best from Spain.

The military establishment of China, including cavalry and infantry, consists of 1,800,000 men. Vide the Appendix to Sir G. Staunton's Account of the Embassy to China.

THE PUNISHMENT OF THE BASTINADO,

Is frequently used in China, for slight offences, and occasionally inflicted on all ranks.

When the number of blows sentenced by the Mandarin are few, it is considered as a gentle chastisement or fatherly correction, and when given in this mild way is not disgraceful, though the culprit is obliged, on his knees, with his forehead touching the ground, to thank the magistrate who so kindly ordered it to be administered.

Every Mandarin whose degree of nobility does not exceed the blue ball on his cap, is subject to this castigation, when ordered by his superior; but all above that rank can only be bastinadoed at the command of the Emperor.

The instrument used on these occasions is a split bamboo, several feet long, which is applied on the posteriors, and, in crimes of magnitude, with much severity. In petty offences, the offender (if he has the means) contrives dexterously to bribe the executioner, who, in proportion to the extent of the reward, mitigates the violence of the punishment, by laying the strokes on lightly, though with a feigned strength, to deceive the Mandarin; and it is said, that, for a douceur, some are ready to receive the punishment intended for the culprit; though, when eighty or a hundred blows is the sentence, it sometimes affects the life of the wretched criminal.

When a Mandarin is from home, he is generally attended by an officer of police, and perhaps one or more soldiers, who are ordered in this summary way to administer some half dozen blows on any careless person who might negligently omit the customary salute of dismounting his horse, or kneeling in the road before the great man as he passes by.

中 国 服 饰

A PAI-LOU, OR TRIUMPHAL ARCH.

These monuments are erected for the purpose of transmitting the meritorious actions of good men to posterity. Magistrates who have executed the duties of their high office with justice and integrity; heroes who have signalized themselves in the field; and others of meaner station whose virtues or superior learning intitle them thereto, often receive this high honour, which likewise serves the purpose of exciting their posterity to the same virtuous actions.

These Pai-lous (usually translated, triumphal arches) are built at the public expense, generally with stone, though sometimes the better sort are made of marble, and some inferior ones of wood; the chief of them have four uprights, each of one stone, which is often thirty feet in length; horizontally across these are placed the transoms or friezes, on which the inscription is engraved with letters of gold, &c. and the summit of the fabric is crowned with projecting roofs richly ornamented.

This was drawn from one near the city of Ning-po, Nov. 17, 1793, where many others are erected, some of which were of a meaner kind, and had but two uprights. The inscription on this was thus translated by a Chinese attendant on the Embassy: "By the Emperor's supreme goodness, in the 59th year of Tchien-Lung, and on the first day, this triumphal edifice was erected in honour of Tchoung-ga-chung, the most high and learned Doctor of the Empire, and one of the Mandarins of the Tribunal of Arms."

VESSELS PASSING THROUGH A SLUICE.

The imperial, or grand canal of China, extends, with little interruption, from Canton, in lat. about 23° 15', to Pekin in 39° 50'.

From this main trunk issue many branches, which pass through innumerable cities, towns, and villages, as roads through European countries; and by this means a communication is kept up with the utmost limits of the Empire; some lesser canals are also cut to counteract the overwhelming effects of inundation; these at the same time serve to convey superfluous water over the low lands for the nutriment of rice, which requires immersion in water till it approaches maturity.

Locks and sluices of various kinds are therefore very numerous; the Print exhibits one chiefly designed as a bridge for the accommodation of foot passengers; the building on the right hand serves to shelter those who are employed in raising the bridge, as well as to preserve the stone under it, which records the name, &c. of the individual who was at the expense of its erection.

Some sluices are so constructed as to retain a considerable body of water for the use of vessels of greater draught; these have grooves cut in the masonry at the opposite piers, into which strong and heavy boards are dropped, similar to a portcullis, and when a sufficient quantity of water is collected, the planks are drawn up and the vessels pass through with considerable velocity, having previously paid a small toll for their admission through the sluice.

The vessel having the yellow or royal flag, is one inhabited by a part of the Embassy; some others occupied by the English have already passed through.

A MANDARIN

attended by a Domestic.

Though chairs are commonly used in China, yet the Chinese sometimes choose to sit in the manner of the Turks.

This Mandarin, habited in his court attire, is one of the literati, and a civil magistrate, which is known by the bird embroidered in the badge on his breast: his high rank and honour are likewise denoted by the red ball and peacock's feather with three eyes attached to his cap, as also by the beads of pearl and coral appending from his neck; he is sitting in form on a cushion, smoaking, and waiting the arrival of a visitor.

The servant bears in his hand a purse containing tobacco for his master; his girdle encloses a handkerchief, and from which also hangs his tobacco pouch and pipe. On the walls of the apartment Chinese characters are painted, signifying moral precepts.

78 中国服饰

A SMALL IDOL TEMPLE,

commonly called a Joss House.

THE general religion of China, Paganism, generates the grossest superstition and credulity among the unenlightended part of the people, who attribute every casual occurrence to the influence of some good or ill star; if the event forebode evil, they immediately repair to the proper idol with offerings, that the impending misfortune may be averted; if good, they also make sacrifices and return thanks.

These sacred edifices are commonly situated near the road side, or on the banks of canals for the convenience of travellers, &c. who are often observed prostrating before them; some are erected at the public expence, and dedicated to former Emperors, Mandarins, and others, for services rendered to their country; and some are built by charitable persons, to extend religious worship among the people.

On days of general rejoicing, as the commencement of the new year, new moon, Emperor ploughing the ground, feast of lanterns, &c. these buildings are much frequented, the people offering before the little gilt images inhabiting the fabric sacrifices of ready dressed animal food, fish, rice, and wine, in proportion to their ability or inclination; while innumerable crackers are fired, and a profusion of gilt paper and incense is burnt before the idol.

Sometimes a priest attends on such occasions to receive these offerings for the benefit of his fraternity, though more frequently the sacrifices of each suppliant are taken to his family and eaten as a feast. The buildings in the back ground are the residence of a Mandarin, known by the two flag staffs at the entrance: on the hill is a military station and a mutilated Pagoda, these being generally erected on an eminence.

CHINESE GAMBLERS

with Fighting Quails.

It is more common in China to breed quails for fighting, than to bring up game-cocks, for the same purpose, in Europe. The male quails, descended from a good stock, are trained with great care; their owners teaching them to fight most furiously, and with a spirit equalling the best of our game-cocks. These battles, though forbidden by the laws, are countenanced and even practised by the Mandarins; and it is a favourite diversion among the eunuchs in attendance at the palace, who often hazard large sums in bets on the issue of a contest. If during a conflict between these little furies, both birds should happen to fall together, that which last endeavours to peck at his adversary, is deemed the victor.

It is said, that oftentimes on the result of these battles, not only the fortune, but even the wives and children of the parties wagering, are put to the chance of being given up to the winner as concubines and servants.

The figure smoking, holds in his hand some Chinese money threaded on a string; the man with a feather behind his cap is betting with him.

中 国 服 饰

PORTRAITS OF SEA VESSELS,

generally called Junks.

On the 5th of August, 1793, the Embassador and his suite left the Lion and Hindostan, and embarked on board the brigs Clarence, Jackall, and Endeavour, when they immediately sailed for the Pay-ho, or White River, in the Gulph of Pe-tchi-li: the other persons attached to the Embassy followed in Junks engaged for that purpose. These vessels, which also conveyed the presents for the Emperor, baggage, &c. are clumsily constructed, and carry about two hundred tons; nevertheless, being flat-bottomed, they draw but little water, and are thereby enabled to cross the shallows at the entrances of the Chinese rivers.

These Junks are of the same form at stem and stern, and the hold is divided into compartments, each being water-tight: the masts are of one tree, and very large; their main and fore sails are of matting, composed of split bamboos and reeds interwoven together; the mizen sails are of nankeen cloth.

The rudders, (which are generally lifted out of the water when at anchor,) are rudely formed, and cannot be worked with dexterity; the steering compasses are placed near them, and surrounded with perfumed matches.

The anchor of four points is of iron, the other of wood; at the quarters are stowed some bamboo spars; and these junks are gaudily adorned with ensigns, vanes, &c. agreeably to the Chinese taste.

84 中 国 服 饰

A SOLDIER OF CHU-SAN,

Armed with a Matchlock Gun, &c.

THE Chinese are supposed to have known the use of fire-arms and gun-powder at a very early period, but since the conquest of that country by the Tartars, the chief expenditure of gunpowder has been in the frequent practice of firing salutes and discharging of fireworks: in the ingenious contrivance of the latter they are eminently skilful.

The army of China is at present very ill disciplined; its strength consists only in its numbers, which would not compensate in the day of battle for their ignorance of military tactics, and want of personal courage.

The general dress of the soldiery is cumbrous, and for the southern provinces almost suffocating, being lined and quilted. At the right side of this figure hangs his cartouch-box, and on the left his sword, with the point forwards. The matchlock is of the rudest workmanship, and has a forked rest near the muzzle.

It must be thought extraordinary that the Chinese government should continue the use of this clumsy weapon, when the ingenuity of the people so well enables them to manufacture muskets equal to those of Europe.

In the back-ground is a military post, having the usual number of soldiers attending it; these are called out by the centinel on the tower, who is beating a gong, to announce the approach of a man of rank, who is entitled to the compliment of a military salute.

EXAMINATION OF A CULPRIT

Before a Mandarin.

THIS subject represents a Female, charged with prostitution. Such an offender is generally punished publicly, by numerous blows with the pan-tsee, or bamboo; and, in cases of notorious infamy, is doomed to suffer the additional sentence of bearing the can-gue; sometimes, however, corporal punishment is commuted into a pecuniary fine.

The Magistrate, habited in full dress, is known to be of royal blood, by the circular badge on his breast, that worn by every other Mandarin being square. The Secretary, who is taking minutes of the proceedings, wears on his girdle his handkerchief and purses, together with a case containing his knife and chopsticks. These purses are merely for ornament, not being made to open.

The Chinese write with a hair pencil and Indian ink: the pencil is held vertically, and the letters are arranged in perpendicular lines from the top of the page to the bottom, beginning at the right and ending on the left side of the paper. The cap worn by the officer of police is distinguished by certain letters which denote the name of the Mandarin he serves.

The manner in which the prisoner is presented is characteristic of the insolence of office and harshness which (even female) delinquents are subject to in that country.

VIEW AT YANG-TCHEOU,

In the Province of Che-kian.

THE city of Yang-tcheou (through which the Embassy passed on the 4th of November, 1793), is of the second order, which is known by its termination, *tcheou.*

The chief building in this subject is a sacred Temple, having the two characteristic flags: on the right is seen a monument, a fort, and part of the city walls.

Chinese fortifications are generally constructed in a manner which Europeans would not consider formidable, but they are, nevertheless, proportional to the efforts of the probable assailants, it being more likely they would be employed against the natives in civil warfare, than against a foreign enemy.

On the fore-ground is seen a tower, and another part of the walls. These defences are in some places continued without interruption over the rivers and canals, and thus become fortified bridges. On the last-mentioned tower and wall are soldiers presenting their shields in front of the embrasures, in compliment to the Embassador. This singular mode of salute, when continued along an extensive line of wall, produced an interesting effect.

On the river are seen many travelling vessels, &c.; the nearest was occupied by a Mandarin attending the Embassy.

TEMPORARY BUILDING AT TIEN-SIN,

Erected for the Reception of the Embassador.

ON the 13th of October, 1793, the Embassy reached Tien-sin, being then on its route towards Canton.

This building of mats (on the banks of the Un-leang), was constructed by order of the chief Mandarin of the city, for the purpose of complimenting the Embassador, and entertaining him and his suite with refreshments, &c.

The landing-place was decorated with mats, fancifully painted; the chief Magistrate of the district sat in a chair, while the inferior Mandarins stood in a rank on each side to receive his Lordship, had he thought proper to debark.

The entertainment consisted of a profusion of poultry, confectionary, fresh fruits, preserves, jars of wine, &c. &c. all which were distributed among the various barges of the Embassy, which are distinguished by their yellow flags.

92 中国服饰

A TRADESMAN.

The dress worn by this person is common among the middle class of the people. The jacket without sleeves is of silk, having a collar made from slips of velvet; the stockings are of cotton quilted, with a border of the same, and his shoes are embroidered.

His pipe, pouch, knife, and chopsticks are suspended from a sash; in his right hand is a basket of birds' nests, which he carries for sale to the epicures of China.

These nests are constructed by birds of the swallow kind, and appear to be composed of the fine filaments of certain sea-weeds, cemented together with a gelatinous substance collected from the rocks and stones on the sea-shore. They are chiefly found in caverns on the islands near the Straits of Sunda, and on an extensive cluster of rocks and islands, called the Paracels, on the coast of Cochin-China.

These nests, when dissolved in water, become a thick jelly, which to a Chinese taste has a most delicious flavour, and communicates, in their opinion, an agreeable taste to whatever food it is combined with. They are therefore highly prized by the upper ranks, and their great expence excludes their use among the poor.

On the bank near which he stands, is a post to which a lantern is attached; the back ground is a scene at Han-tcheou-foo.

A FUNERAL PROCESSION.

The leader of this solemn pageant is a priest, who carries a lighted match, with tin-foil and crackers, to which he sets fire when passing a temple or other building for sacred purposes. Four musicians with gongs, flutes, and trumpets follow next; then comes two persons with banners of variegated silk, on the tops of which two lanterns are suspended; these are followed by two mourners clad in loose gowns, and caps of coarse canvas; next to these is the nearest relative, overwhelmed with grief, dressed in the same humble garments, and is prevented from tearing his dishevelled hair by two supporters, who affect to have much ado to keep the frantic mourner from laying violent hands on himself; then follows the corpse, in an uncovered coffin, of very thick wood varnished, on which a tray is placed, containing some viands as offerings; over the coffin is a gay ornamented canopy carried by four men; and lastly, in an open carriage, three females with dejected countenances, arrayed in white, their hair loose, and fillets across their foreheads.

Contrary to European ideas, which comsider white as the symbol of joy, and use it at nuptial celebrations, it is in China the emblem of mourning, and expressive of sorrow.

The scene is at Macao: in the fore ground is a large stone with a monumental inscription; in the distance is seen the inner harbour, and the flag staves of a bonzes' temple.

A STONE BUILDING

in the Form of a Vessel.

In one of the courts of the hotel, appointed for the residence of the Embassador in Pekin, was an edifice representing a covered barge; the hull was of hewn stone, situated in a hollow or pond that was filled with water, which was supplied from time to time by buckets from a neighbouring well, as might be necessary; the upper part of this whimsical building was used by part of the suite of the Embassy as a dining room.

The fragments of rocks artificially piled on each other with flowerpots, containing dwarf trees here and there interspersed, will convey in some degree an idea of Chinese taste in ornamental gardening on a small scale.

Over the roof of the stone vessel, and above the wall enclosing this extensive mansion, the tops of a few pagodas, a triumphal arch, and other public buildings were visible within the walls of the great city of Pekin.

This great mansion was built by a late (Hoppo or) collector of customs at Canton, from which situation he was promoted to the collectorship of salt duties at Tien-sien; but his frauds and extortions being here detected, the whole of his immense wealth was confiscated to the crown.

A FISHERMAN AND HIS FAMILY,

regaling in their Boat.

The female of the group, surrounded by her children, is smoking her pipe. One of these has a gourd fastened to its shoulders, intended to preserve it from drowning, in the event of its falling overboard.

The whole family sleep under the circular mats, which also serve as a cover to retreat to in bad weather; through the roof is a pole, surmounted by a lantern, and on the flag are depicted some Chinese characters.

On the gunwale are three of the leutze, or fishing corvorants of China; in size, they are nearly as large as the goose, and are very strongly formed both in their beak, their legs, and webbed feet. On the lakes of China, immense numbers of rafts and small boats are frequently seen employed in this kind of fishery. A well-trained bird, at a signal from its master, immediately plunges into the water, and soon returns with its prey to the boat to which it belongs; sometimes it encounters a larger fish than it can well manage, in which case the owner goes to assist in the capture; it is said indeed, that these birds have the sagacity to help each other.

That the young leutzes may not gorge their prey, a ring is put on their neck to prevent its passing into the stomach; when they have taken enough to satisfy their master the ring is taken off, and they are then allowed to fish for themselves.

Beyond the boat is a sluice, or flood-gate, for the passage of vessels. The distances behind indicate the serpentine direction of the canal.

INDEX.

Dedication Plate to follow the Title-page.

No. 1.
- Portrait of Van-ta-zin.
- A Peasant, with his Wife and Family.
- A Pagoda, or Tower.
- The Travelling Barge of Van-ta-zin.

No. 2.
- A Chinese Soldier of Infantry, or Tiger of War.
- A Group of Trackers at Dinner.
- View of a Bridge at Sou-tcheou.
- Portrait of a Trading Ship.

No. 3.
- Portrait of the Purveyor of the Embassy.
- Punishment of the Can-gue.
- South Gate of the City of Ting-hai.
- Three Vessels lying at Anchor at Ning-po.

No. 4.
- Portrait of a Lama, or Bonze.
- A Chinese Lady and her Son.
- View of a Burying-place.
- Front View of a Boat passing over an inclined Plane.

No. 5.
- Portrait of a Soldier in his full Uniform.
- A Group of Peasantry, Watermen, &c.
- View of a Castle near the City of Tien-sin.
- A Sea Vessel under Sail.

No. 6.
- Portrait of Van-ta-zin in his Dress of Ceremony.
- A Chinese Porter or Carrier.
- The Habitation of a Mandarin.
- A Mandarin's travelling Boat.

INDEX.

No. 7.
- A Standard Bearer.
- A Sacrifice at the Temple.
- A Military Station.
- A Fishing Boat.

No. 8.
- A Chinese Comedian.
- A Group of Chinese, habited for rainy Weather.
- A Pagoda, or Temple for religious Worship.
- A Ship of War.

No. 9.
- A Soldier in undress, with a Flag at his Back.
- The Punishment of the Bastinado.
- A Pai-loo, or Triumphal Arch.
- Vessels passing through a Sluice.

No. 10.
- A Mandarin, attended by a Domestic.
- A small Idol Temple, or Joss-house.
- Chinese Gamblers with fighting Quails.
- Portrait of Sea Vessels generally called Junks.

No. 11.
- A Soldier with a Matchlock.
- A Criminal brought before a Magistrate.
- Suburbs of a City, Canal, &c.
- Temporary Building at Tien-sin.

No. 12.
- A Tradesman with Birds' Nests for Sale.
- A Funeral Procession.
- Building, resembling a Vessel.
- Fisherman and his Family in a Boat.

Printed by W. Bulmer and Co.
Cleveland-row, St. James's.

图书在版编目（CIP）数据

中国服饰 /（英）威廉·亚历山大著；徐锦华主编.
-- 上海：上海古籍出版社，2023.6
（徐家汇藏书楼珍稀文献选刊）
ISBN 978-7-5732-0613-8
Ⅰ.①中… Ⅱ.①威… ②徐… Ⅲ.①服饰文化—
中国 Ⅳ.①TS941.12
中国国家版本馆CIP数据核字（2023）第032053号

丛书主编：徐锦华
丛书总序：董少新
本册导言：陆丹妮

责任编辑：虞桑玲
装帧设计：严克勤
技术编辑：隗婷婷

徐家汇藏书楼珍稀文献选刊
中国服饰
The Costume of China

［英］威廉·亚历山大（William Alexander）著

上海古籍出版社出版发行
（上海市闵行区号景路159弄1-5号A座5F　邮政编码201101）
　（1）网址: www.guji.com.cn
　（2）E-mail: guji@guji.com.cn
　（3）易文网网址: www.ewen.co

印刷：上海丽佳制版印刷有限公司

开本：787×1092毫米　1/8
插页：5　印张：14.5　　字数：112千字
版次：2023年6月第1版　2023年6月第1次印刷
ISBN 978-7-5732-0613-8 / J·673
定价：358.00元